Science Fictioned

Volume 1

Lee Falin, PhD

LIGHT & LORE LLC

Also by Lee Falin

The Science Fictioned Series

Science Fictioned - Volume 1

Science Fictioned - Volume 2

Standalone Novels

Half Worlder

Science Fictioned
Volume 1

About Science Fictioned

Science Fictioned takes ideas from cutting-edge, scientific research papers and turns them into science fiction and fantasy stories. This book has its roots in two observations I made while working as a researcher in Europe:

1. There are some really amazing scientific discoveries being made in the world these days.
2. Research papers are so boring, that most people are probably never going to hear about those amazing discoveries unless they happen to make the 6 o'clock news.

I was reading over the details of yet another extremely dull paper one day, trying to keep my colleagues from noticing the drool on my chin, when I realized that there was a better way— what if every research paper was accompanied by a thrilling story?

In classic science fiction stories, authors like Asimov and Heinlein took ideas about science and built wonderful stories around them. What if journal publishers started requiring scientists to write entertaining stories about the research papers they were submitting?

I'm not talking about those humorous anecdotes you often hear in seminars regarding the time a shockingly underpaid lab assistant accidentally mixed up a set of tissue samples stuck in a liquid nitrogen bath, which then took hours to sort out properly (a purely hypothetical example). I'm talking about actual story-telling in the form of classic science fiction and fantasy short stories.

I knew at once that this idea was nothing short of pure genius. Scientists would get their research out there, the public would have a better understanding of cutting-edge research, and scientific journals would sell like hotcakes. Everybody wins.

I immediately took the next logical step: I quit my job as a lowly researcher and headed off to Washington to sell my idea to the top brass. I met with men and women from the NIH, the USDA, the CIA, and the IRS (that last meeting wasn't my choice, but that's a different story).

Everyone I spoke with was impressed with my idea, and I was pretty sure once the administrative assistants and security guards forwarded my requests to the right people, things would start to change. It was an exciting time to be alive.

I was strolling down Army Navy Drive, eating a cone of choco-cherry ice cream, wondering how I could sneak into a meeting with the Joint Chiefs of Staff, when I ran into the editor of a very prestigious scientific journal.

Given my mission to reshape the world of scientific litera-ture, this would have been rather fortuitous had I not, quite literally, run into him. As I hurriedly expressed my apologies and tried unsuccessfully to use his tie to wipe choco-cherry ice cream off his Armani suit, I gave him the elevator pitch version of my brilliant idea.

Before I was even halfway finished, the editor held up a hand to stop me. "I'm sorry son, it'll never work."

I stared at him, dumbstruck, every single secretary and security guard I'd spoken with today had thought my idea was brilliant, what was this guy's problem? Maybe he didn't like choco-cherry ice cream?

"The grant committees will never go for it," he said sadly.

With a feeling of despair, I realized he was absolutely right. You might not be aware of this, fair reader, but almost all scientific research these days is funded by grants. Scientific grants aren't like those grants you might have gotten to attend school, where the government just throws money at you and sends you on your way with a nod and wink.

To get a research grant, a scientist has to write a grant proposal in which they outline exactly what they want to do, why they are the best ones for the job, and how much they think everything will cost. Writing a grant proposal is often such a difficult job that an entire team of scientists will band together to tackle it. The process can take months of research, writing, editing, luncheons, faculty retreats to Tahiti, and other onerous duties.

Once the grant proposal has been submitted, a review committee has to approve it. They weigh in on the merits of the idea, the qualifications of the researchers, and the relative insanity of the budget. Every single item in the budget is scrutinized to ensure that public money isn't wasted on anything frivolous—at least, nothing the committee thinks is frivolous.

"Grant committees," I muttered as the last of my choco-cherry ice cream dribbled out of my cone. "I had forgotten."

"Lucky you," the editor muttered. "But the fact is, no grant committee would ever sanction paying for something as frivolous as a science fiction story."

"Yes, yes," I muttered. "You're quite right, of course...I

appreciate your feedback." With somewhat less zeal than before, I resumed cleaning ice cream from his suit.

"Don't worry about it, son," he said, sidestepping my half-hearted attempts to pull a piece of cherry out of one of his buttonholes. "I'm on my way to a research conference in Tahiti and there's a generous allotment in the grant for dry cleaning." He gave me a quick nod, a sympathetic smile, and hurried away.

I was left alone with an empty ice cream cone and a broken dream. I was now faced with a choice. I could either let my dream die right there on a lonely sidewalk in Washington, or I could go it alone, write the stories myself, on my own dime.

And thus, Science Fictioned - Volume 1 was born: an entertaining collection of science fiction and fantasy short stories based on ideas from cutting-edge, peer-reviewed, scientific research articles.

If you're reading this and are part of a grant review committee, I hope you pay extra-close attention.

Part One
Mice, Bacteria, and Goblins

The Science

B acteria are a lot smarter than most people give them credit for. Granted, they do spend most of their short lifespans moving around aimlessly in search of good things to eat, responding to chemical signals from other members of their species, and occasionally ganging up on weaker creatures they encounter. Humans are obviously *way* more advanced than that.

It's common to think of bacteria as those nasty little buggers that make us sick when we accidentally forget to wash our hands before eating. However, many strains of bacteria floating around in the world don't make you sick at all. In fact, some of them can be quite beneficial to humans.

Bacteria that cause illness and disease are called pathogenic —path-OH-jinn-ick—bacteria. This term comes from the Greek root *pathos*, meaning "suffering" and the Arabic word *djinn*, meaning a magical sprit. So, you can think of pathogenic bacteria as magical creatures that make you suffer.

Editor's Note: Approximately 76% of the word origins in this book are completely made up. We tried to get a straight answer from the author as to which ones could be relied upon, but he refused to cooperate. It is therefore left as an exercise to the reader to discover which word origins are true, and which are completely fictitious. We strongly advise against using this book as an SAT prep guide.

Pathogenic bacteria usually need to spend at least part of their lifecycle inside of a host. Depending on the particular type of bacteria, that lucky host could be a flower, a rat, or your Uncle Larry. When pathogenic bacteria first invade a host, they could attack right away. But if they act too soon, the host's immune system will usually snuff them out quickly.

In the United States Congress, important matters aren't supposed to be voted upon unless there are enough representatives present. When there are enough of them in the room to vote on something, this is called a "quorum". This term comes from the Latin root *qui* meaning "what" and the old English *rum* meaning "room". You can think of a quorum as a bunch of people wondering what room they're supposed to go to in order to vote.

When Congress is about to vote on an issue, someone issues what is called a "quorum call" where they check that there are enough legislators around to call for a vote. This is a lot like roll call in primary school. If a congressman isn't there, he or she needs to bring a note in from their mother explaining why they missed the vote.

Congress didn't come up with this quorum idea all by themselves, they got the idea from bacteria. Before bacteria vote on whether or not to attack the host they're living in, they use a process called "quorum sensing" to detect if there are enough of them around to survive the attack.

But because bacteria can't talk, they can't use the same roll call method that Congress uses. Instead, as each bacterium floats around, it generates little signal molecules to indicate its presence. Whenever enough of these signal molecules build up in the host, the bacteria decide there must be enough of them around to start the attack.

Scientists have known about this process for quite some time, but it took them a while to figure out just how to write the appropriate grant proposal to do anything about it. Finally, in 2010, a pair of intrepid scientists from Cornell named Duan and March received a grant from none other than Bill Gates (yes, *that* Bill Gates) to investigate the problem.

In an effort to trick the bacteria into attacking too soon, Duan and March developed a way to take a harmless strain of bacteria that give off the same quorum-sensing molecule as cholera bacteria (which are typically not so harmless). They figured they could use the harmless strain of bacteria as a sort of false alarm, a double agent that would trick the cholera into attacking too soon.

An interesting fact about university science labs is that they are almost all underground. This is because the athletic department gets first pick of university facilities, followed by the business and engineering departments. That leaves the poor science departments to fight over whatever space is left over.

Because of this method of real estate allocation, most university science labs end up in the basements of whichever buildings have the worst heating. Perhaps not coincidentally, these same buildings tend to have rather serious rodent infestations. Fortunately, since getting brand new equipment and facilities requires the approval of a grant committee, most scientists are quite adept at making do with whatever resources they can find lying around.

This is why scientists started using mice in their experiments; not because there is anything particularly scientific about mice, but simply because mice are always underfoot in the dark and musty basement labs where scientists are often forced to work.

So, it's no surprise that our friends Duan and March, working in their cold and drafty basement lab over at Cornell, decided to use mice in their quorum sensing experiments.

First, they took some delicious mouse food they'd picked up at the local supermarket, mixed it with some plain bacteria, then fed it to a group of mice. Just as they'd gained the trust of their rodent companions, the scientists injected them all with cholera.

The second group was rather tremulous after witnessing this. They were hoping they'd somehow get off easy, and at first things seemed to be going well. No weird food supplements were brought out. Perhaps the scientists were done with mice and were just going to let them go. But then the scientists injected this second group was cholera too.

The third group of mice were just about to make a break for it, when they saw the scientists take out a suspicious-looking jar filled with powder. They mixed liberal quantities of this powder with the mouse food and placed it in the mice's feeding bowls. The label on the jar said "harmless, genetically engineered, double-agent bacteria".

Even though the mice didn't trust the scientists, they weren't about to pass up a free meal, even if it was laced with genetically engineered, double agent bacteria. After all, the label said it was harmless, so they devoured the food, and were feeling quite smug about the whole thing until a few hours later when the scientists came along and proceeded to inject them with cholera too.

By this time, all the mice were rather incensed by this treatment. But all things considered, there didn't seem to be any lasting side-effects. They were a little sore perhaps, and one or two had a stomachache, but that was to be expected after eating so much. Perhaps life in the labs wasn't as bad as they'd always heard. They kept reassuring each other about this until the scientists started directing them.

Had the mice listened to their mothers, they would have realized that the scientists weren't interested in them at all. They were interested in how the cholera was getting on. Unfortunately for the mice, the cholera bacteria were living in their intestines.

In the first group of mice, which had received plain bacteria plus cholera, the scientists found the cholera going along, wreaking havoc in the mice's intestinal tracts, just as they had expected. In the second group, which had received just a cholera injection, things were just the same.

However, it was the final group of mouse intestines that really excited those scientists. The cholera in the third group of mice, the ones that had received the special double-agent bacteria plus the cholera injection, had considerably less cholera running around than the first two groups.

Duan and March concluded that this was because the signal molecules given off by the double-agent bacteria had tricked the cholera into attacking too soon. This gave the mice's immune system an advantage in fighting off the cholera. The scientists published their paper with these results, giving no credit whatsoever to the mice. The grant review committee was impressed, and our two clever scientists are now living the good life, wearing Armani suits, and attending research conferences in Tahiti.

However, what neither Duan and March, their mice, nor the grant review committee knew was that like many great discoveries in science, quorum sensing is a concept that farmers have known about since ancient times. This isn't something discussed by modern farmers, but it was this very principle that led to the widespread use of the scarecrow.

The Fiction

—— ✧ ——

G rog smiled to himself as he glanced up at the crescent moon. It was the perfect night for a hunt. There was just enough moonlight for a goblin to see by. They'd definitely have the advantage over the human scum this time. A light breeze rustled the corn stalks that surrounded him on all sides.

He couldn't see the farmhouse from here, not with so many stalks of corn, but he knew it wasn't far. If all went well, by this time tomorrow he and his clan would be dining on fresh—a loud snap behind him caused Grog to spin around.

He was just in time to see Smurg scramble to his feet, looking embarrassed.

"Quiet!" Grog hissed, glancing in the direction of the farmhouse. It was unlikely anyone had heard them at this distance, but you couldn't be too careful. Some humans were clever.

"It's not my fault," Smurg muttered as he squeezed noisily through two stalks of corn to join Grog. "It's dark and this ground is uneven."

Grog rolled his eyes. "Dark," he muttered. "What kind of goblin hunter can't see in the dark? That's what I want to know." Not for the first time, Grog wondered which of the seven cursed spirits was responsible for giving him the clumsiest goblin in the clan for a brother.

Smurg didn't answer. He was squinting through the stalks of corn in the direction of the farmhouse. "Looks empty," he said. "All the lights are out."

"They're asleep, bone-brain," Grog growled. "Now be quiet or they'll wake up and we'll miss another meal."

Smurg stood quietly for a minute, then glanced at Grog.

Grog ignored him.

Smurg opened his mouth like he was going to speak, then shut it again.

Grog continued to ignore him.

Smurg looked back towards the cottage, then glanced yet again at Grog.

"What?" Grog snapped.

"Shouldn't we move in closer?"

Grog fought to keep the growl out of his voice. "Good idea Smurg," he said. "Why don't you run on ahead and try sneaking through the kitchen window like you did last time."

Smurg winced. "That wasn't my fault. That window was smaller than it looked."

Grog smirked. "Right."

"Mom says I'm just big boned," Smurg sniffed. "She says all the bravest goblin warriors had big bones."

"Unfortunately, your biggest bone is the one in your head. Now be quiet so we don't miss the signal."

Smurg made a loud hooting noise.

Grog spun around and stared at him, opened mouth. "What —in the name of Grabthar's hammer are you doing?"

Smurg cowered slightly. "Mom said you shouldn't use language like—"

"If you do that again," Grog said menacingly, "the next thing you'll hear Mom say will be Fragnar's funeral chant."

"I was just practicing the signal."

"That's not the signal toad-face. We're supposed to wait quietly until we see Brock wave his arms towards the cottage. Then we move in—silently."

Smurg's upper lip drew down into a pout. His eyes scrunched together, and tears started to form in their corners.

Grog held up his claws in alarm. "No! Don't—"

Smurg let out a wail. "I-I d-d-don't like it when you c-c-call me n-names," he wailed.

"If you don't stay quiet," Grog growled, glancing around hurriedly, "I'm going to do a lot worse than just call you names."

Smurg sniffed again as he wiped his nose on his ragged tunic. He still looked miserable, but at least he'd stopped wailing.

"We're still early," Grog muttered, glancing up at the sliver of moon. "Brock said he'd be in position before the moon hit its zenith. You better hope that he didn't hear your wailing, or this will be the last hunting party you go on. No matter what mom—"

"There he is!" Smurg said, jumping up and down as he pointed off to their left.

Grog looked where he was pointing and saw a dark shape standing near the center of the field. Its arms were extended towards the cottage.

Smurg squinted towards the figure. "Is he making the signal?"

"I don't—" Grog broke off as the wind shifted. "Doesn't smell like Brock," he muttered. He glanced at the moon again. It seemed too early.

The wind was picking up now, making the corn rustle noisily.

"He moved his arms!"

"What? I didn't see—"

"He must have been waiting for the wind, to give us more cover," Smurg said, fingering his club nervously.

"Don't you think I know that?" Grog snapped. He couldn't stand it when Smurg pointed out something that he had missed. "Now stop playing with your club and let's move."

He crouched low and started creeping through the field, quieter than a mouse. His parents always said he had a gift for stalking, he—

"Grog?" Smurg called in a half-whisper from behind him. "Where are you?"

Muttering words he would never use in front of his mother, Grog retraced his steps until he was behind Smurg. "I'm right here numbskull," he whispered.

Smurg yelped and spun around. "Don't sneak up on me!"

Grog grabbed him by the tunic and pulled him down next to him. "Follow me," he growled. "And be quiet." He glared at Smurg until his brother swallowed and nodded. Then he let go of his tunic and stalked silently back towards the cottage. His brother followed in a half-crouch, trampling the stalks of corn behind him.

They were nearly to the edge of the cornfield now. He could see the outline of the farmhouse windows. He squinted to his left, trying to see if Brock and the others were in position yet. If he attacked too soon—

Thwack!

Grog dove backwards into deeper cover. That sound was unmistakable. Cursed humans. He turned to warn Smurg to—

thwack!

Smurg stared down at the arrow protruding from his chest. He touched the arrow's fletching and then looked up at Grog, his eyes wide with surprise. Then he fell to the ground, lifeless.

15

Grog stared at his brother's fallen form. He was surprised to see that he had reached one claw towards Smurg as if trying to catch him when he fell. He let his arm drop slowly to his side. Moving quickly, he crawled back to where his brother lay and knelt over him. He quietly muttered Fragnar's funeral chant. Then passed his clawed hand over his brother's eyes, closing them forever.

He felt a lump in his throat as he imagined his brother's soul clumsily entering the eternal goblin hunting grounds. Shame washed over him as he remembered scolding his brother earlier. *The next thing you'll hear Mom say will be Fragnar's funeral chant.*

What would Grog tell his mother? Smurg had always been her favorite. If he came back without his brother, what would it do to her? Grog's eyes narrowed as he turned back towards the cottage. The only thing he could do for his brother now was avenge him. The humans would pay this night. They would learn the folly of trying to fight back against a full goblin hunting party.

Confident that Brock and the others were in position by now, he lifted his head and let loose the one sound that travelers feared above all others. The sound that sent the bravest of warriors into hiding—the goblin death charge.

His rage fueled him as he ran. He felt unstoppable as he ran snarling towards the building, blood pumping in his ears. He barely slowed as he reached the clearing surrounding the farmhouse. Raising his club to smash the nearest window, he screamed in triumph.

"For Smurg!"

Thwack!

An arrow slammed into his club, sending it spinning out of his claws. He turned with a growl and saw a human peering through the upper window of the farmhouse, its fleshy hands nocking another arrow to his bow.

"There! In the window. He—", Grog broke off as he noticed for the first time the silence that surrounded the cottage. He was alone.

Where was the rest of the hunting party? Surely, they hadn't *all* been killed. He looked back towards the part of the field where he had first seen Brock, then felt his blood go cold. It couldn't be...Brock was standing in the same spot, his arms still outstretched.

A cloud shifted away from the moon, and Grog could make out the face on the figure. It wasn't Brock's face, it wasn't even a goblin's face. It was the face of a man made of straw—the face of a scarecrow. The last face that Grog would ever see.

Thwack!

~

Thomas yawned and rubbed his eyes as he stepped into the bright sunlight. The sun was already well up in the sky. He wrinkled his nose at the smell that hung in the air. It smelled like someone had dipped a cat in raw sewage and then set it on fire.

Thomas knew he shouldn't have slept so late, but it had been a long night. His father would probably already be working on the new well, he'd better hurry along and help him or he'd really catch it.

He moved towards the barn on the other side of the clearing, quickening his step slightly as passed the green stain in the dirt near the side of the farmhouse. His father must have already disposed of the dead goblins. That would explain the smell.

He looked reflexively towards the goblin decoy his father had insisted on putting up in the field earlier in the Spring. His mother had chided his father for what she had called 'superstitious nonsense'. But even she couldn't deny it had saved their lives.

Thomas shuddered as he remembered the guttural yell the second goblin had made after its companion had been killed.

Luckily there had only been two of them.

Part Two
Zombie Ants and Socialites

The Science

Just when you thought life couldn't get any creepier, one day an entomologist went and discovered zombie ants. The word *entomologist* comes from the Greek root *Entomos*, meaning "to cut up" and the Latin *logia*, which is the root of our modern day "Eulogy", a speech given about someone after they die. So, entomologists are scientists who like to cut things up and then give speeches about what they found, which explains their fascination with insects.

Zombie ants are ants which, when infected with a certain fungus, start behaving like zombies. Not the kind of zombies that hunt people down and try to eat their brains, but the kind that have a single-minded purpose. If you haven't had the pleasure of meeting a zombie ant, which is unlikely unless you're an entomologist or live in a rainforest, allow me to introduce you.

A zombie ant is like a regular ant, but instead of being busily involved in its normal little ant life, it is suddenly seized with a compelling urge to wander around in a sort of drunken stagger until it finds a nice leaf high up in a tree.

It then climbs under the leaf and bites down hard, staying frozen in that position until it dies. The fungus inside the ant

21

then explodes, releasing a cloud of spores that go on to infect other unfortunate ants.

Scientists believe the reason the fungus compels the ants to climb to the underside of a leaf is due to the advantage such a position would give to fungus wanting to spread spores as widely as possible.

Unfortunately, the weirdness doesn't stop there.

A certain type of virus has been found to have similar effects on the gypsy moth caterpillar. Normally a cautious and nocturnal creature, the gypsy moth caterpillar spends most of its daylight hours hiding under loose tree bark. However, once it is infected with a certain virus, the normally sleepy caterpillar climbs as high as it can in the treetops, where it dies and turns to mush. The mushy remains are then scattered by the rain, allowing the virus to find more caterpillars to infect.

Obviously, this kind of behavior begs to be understood. Not just for the sake of the gypsy moths, but to make sure that humans don't fall prey to such antisocial behavioral changes. What if that weird fellow who hangs out on the corner watching traffic pass isn't just some creepy antisocial vagrant? What if he is just one more victim to a behavior-altering virus or fungus?

Fortunately, a couple of talented scientists at Pennsylvania State University decided to dig deeper into this mystery. First, they slipped the grant community some viral samples to alter their behavior enough to approve their research proposal. With that task out of the way, they set out to discover just how the virus carried out its nefarious work on hapless caterpillars.

These scientists decided to put some caterpillars in jars, then infected them with different viruses to see what would happen. Now, a normal person would have injected each caterpillar with the same virus, just to be fair. But scientists aren't fair-minded at all. Disparity in experimentation is our mantra.

So, the scientists gave each caterpillar a virus that was slightly different.

They had previously suspected, for reasons apparent only to entomologists, that a certain gene inside of the virus was the cause of the caterpillar's odd behavior. To test their theory, they cut that gene out of the virus—*entomos*, remember?— then they gave some caterpillars the normal virus, and gave others the virus with the gene cut out.

They discovered the caterpillars who received the normal virus climbed to the top of their jars and died, just like their slightly happier, but still doomed-to-die counterparts in the forest.

The interesting part came when they looked at the caterpillars who got the modified viruses. Those caterpillars climbed up to the top of the jar as well but then climbed back down before they died, which I think we can all agree is considerably more exciting. This led the scientists to conclude that the gene they cut out of the virus was at least partially responsible for affecting the behavior of the caterpillar.

All this zombie-like behavior made me wonder: what if there are viruses and fungi floating around that can influence human behavior? If I were a virus, and I wanted to spread to as many new people as possible, what would I make my host do? Certainly not climb a tree and melt—people tend to avoid things like that.

If I were a virus, I'd want my host to be popular.

The Fiction

$$\star$$

"You are so popular!" Mark gushed for what must have been the tenth time today.

Several kids walking through the quad glanced at us. A couple of them waved. I smiled and waved politely, then went back to trying to organize my backpack.

Mark waved the school paper in front of me, but I didn't bother looking up; I didn't need to. I had the headline memorized:

Record-breaking victory for Susan Mitchell, Carter Middle School's Newest Student Body President.

"So, Mrs. President," he drawled, "what's your first order as our new leader?"

"I'm going to banish annoying fans," I said, pushing the paper away. I didn't mean to sound so harsh, but he was going to put my eye out with that thing.

Mark just laughed it off and started reading through our list of campaign promises.

"We could try for the extended lunch periods thing...or

maybe that one about letting students leave early on Friday afternoons..."

Even though Mark could be annoying sometimes, I had to smile at his enthusiasm. Typical Mark, straight to business. He'd been that way for as long as I'd known him—four solid years of to-do lists and planners. The first time he'd come over to my house, he'd offered to help my Mom organize the pantry.

"Hey, aren't you listening?"

"Huh?" I asked, looking up.

He frowned. "I asked if you thought the vending machine plan would be too much too soon."

"Umm, yeah, maybe. But we can start on it right after I get back."

"When you get back?" he asked, confused. "Where are you going? Your family isn't going on vacation *now,* are they? We've just been elected!" He brandished the paper in the air again. "This is a critical time in your presidency."

I felt a twinge of guilt as I rummaged around in my backpack, trying to avoid his eyes. "Um—didn't you hear the morning announcements yesterday?"

"I was late to homeroom because I was hanging up some last-minute campaign posters in the cafeteria. Why? What'd I miss?"

"Well..." I hesitated, knowing that Mark wasn't going to like this. "Principal Whittaker said that the eighth grade Class Presidents from each school in the state are going to a special lunch with the governor."

Mark's face lit up with excitement. "Wow! How are we getting there? Are they going to rent a limo for us or fly us in—" He stopped abruptly, and his eyes narrowed. "You said Class *Presidents*...so does that mean we lowly Vice Presidents don't get to go?"

"I'm really sorry Mark," I said, forcing myself to look at him.

"I didn't even know about it until this morning. I asked Principal Whittaker if you could come too, but he said the budget for the event only covers one student from each school."

Mark looked away, but not before I saw the hurt in his eyes. "No problem," he said coolly. "I'll just work on some more posters or something."

"Mark don't be like that," I pleaded.

Dad had warned me that something like this might happen. When I told him that Mark and I were planning to campaign together, he'd warned me not to mix relationships and politics. I tried to explain that Mark and I didn't have a relationship, that we were just friends, but Dad still thought it was a bad idea.

Now, seeing the distant look on Mark's face, I wondered if Dad had been right after all. About the campaign that is—not the relationship.

"How about this," I said, "you can pick any of the things from the list we made, whichever is your favorite; and we can get started on it right away."

"*I* can get started right away you mean," he muttered. "You'll be off hobnobbing with the governor."

Despite his tone, I could see his eyes take on that familiar intense look as he scanned over the list. Mark loved lists, and most of the ideas on it had been his anyway.

~

I hugged myself and bounced up and down as the wind picked up. There was no cover from the winter wind in front of the school. I should have worn a heavier jacket.

Principal Whittaker checked his watch again and glanced towards the school entrance. "Any minute now," he said, trying to sound enthusiastic through his chattering teeth.

We were standing a little apart from the crowd of students that had gathered to see their lucky class president picked up in a real limo. I was glad that, despite the cold, the forecast hadn't called for snow today.

I glanced quickly over at Mark who was talking to Cathy Walker from our Algebra class. I could tell from the envious looks he kept shooting me that he was still annoyed at being left out of this trip. Cathy didn't notice though; she seemed to be enjoying the attention. Since when had Mark been interested in Cathy Walker?

I wondered if talking to Cathy was his way of getting back at me. Even though he had told me several times on Friday that he wasn't upset about the trip, he hadn't called me at all over the weekend. I'd been dreading seeing him this morning.

Cathy laughed at something Mark said, and he blushed.

Maybe I didn't care what he thought about the trip. No, I scolded myself, that isn't the right attitude. I decided to follow my mom's advice and try to be the better person. I tried to think of something comforting to say to him, something that would help smooth things over, but my thoughts were interrupted by the arrival of the limo.

"There it is!" someone shouted.

My eyes were drawn to the road as everyone started cheering and talking at once. A long black limo was rolling slowly up the school driveway. It slowed to a stop alongside the curve where Principal Whittaker and I stood.

"Ready, Miss Mitchell?" Principal Whittaker asked with a broad smile.

"Just a second," I said, trying not to sound as nervous as I felt. I looked back at Mark again, he was staring at the limo was a strange look on his face, like he had just swallowed something his stomach didn't agree with.

"Be the better person," I muttered to myself as I walked over to say goodbye to Mark.

His eyes flicked to my face, then back to the limo.

"Um...see you later?"

Mark just kept staring at the limo.

I swallowed and tried again. "When I get back, you can tell me how it's going with those vending machines." I really didn't care if the school got new vending machines next to the cafeteria or not—I didn't even drink soda. Mark had been excited about it though, and so I was trying to sound interested, for his sake. My efforts were wasted though. He didn't look at me once.

I felt a rush of annoyance. If he doesn't even want to bother talking to me, then he—

I lost my train of thought as I noticed the strange look on his face. He had an intense look in his eyes, and his veins had started to bulge in his neck. Beads of sweat streamed down his face. He wasn't dressed any more warmly than I was though, just a sweater and light jacket.

"Mark are you—"

"I'm—fine," he said through gritted teeth. "Have—a good time—okay?"

"Um...okay..." I said, confused.

Was he sick, or was he really *that* angry with me? I felt my eyes sting as I turned away. I would *not* cry in front of the whole school. "It's not like this is my fault," I grumbled as I stalked towards the limo.

I waved at the crowd, and they erupted into a new round of shouts and cheers. Someone yelled, "I love you, Mrs. President!" and I blushed.

As I started to get into the car, several things happened at once. The cheers changed into cries of alarm. Someone grabbed my shoulder. I turned and saw Mark right behind me, a wild look in his eyes, his fingers digging into me.

"Ow! Stop it, Mark. You're hurting me."

Mark made a strange growling sound as the principal tried to pull him back from the car and pry his hand off my shoulder at the same time. A couple of teachers ran over to help as Principal Whittaker barked something into his walkie-talkie.

Mark was staring straight at me with a frenzied look in his eyes as he struggled against the three teachers who were holding him back.

"What's wrong with him?" I sobbed, unable to hold back the tears. Mark was such a jerk. I hated crying in public.

"It's okay dear," said a calm voice. Mrs. Newell, the school nurse, was suddenly there with her arm around me. "Let's go inside out of the cold for a few minutes and make sure you're okay."

She tried to steer me towards the front office, but I resisted. "What's wrong with Mark?" I choked out again.

"He'll be fine dear. Let's just go inside."

I tried to pull away. I didn't want to go inside. I wanted to know what was wrong with Mark. Why had he attacked me? I looked back to see if he was still staring at me, but he wasn't. He was staring at—the limo.

~

I stood alone in Mark's hospital room. Well, I wasn't technically *alone*, since Mark was there too, but seeing as how he'd been unconscious for the past six hours, that didn't really count.

The light from the setting sun streamed in through the window, making eerie patterns on the floor. The monitors connected to Mark beeped in a steady and predictable way. Mark would have liked that.

A nurse came in and checked the monitors. "Visiting hours will be over in about an hour dear," she said without taking her eyes from the monitors. She made some notes on Mark's chart and left the room.

I walked towards the hospital bed, close enough to see Mark's chest rise and fall with his breathing. He looked so peaceful lying there. It was hard to believe that just a few hours ago he was acting so crazy. I reached out to feel his forehead, though I wasn't sure why. Maybe because that was what my mother always did when I told her that I wasn't feeling well. I jerked my hand away as I heard someone enter the room.

I'd expected it to be my mom. She had taken Mark's mother, Mrs. Callahan, down to the cafeteria for a break, and to give me some time alone with Mark. But it was Dr. Patterson, Mark's family doctor. He was with two other doctors that I didn't know. They were all wearing masks and gloves, like they had just come out of surgery.

One of the new doctors stepped forward. "Miss Mitchell," he said, his voice strangely muffled by the surgical mask, "we need to ask you a few questions...about Mark."

I sighed. It felt like I had already answered every possible question about Mark several times over. I had explained it all to the police and paramedics at the school, then to Mark's parents, then my parents, and then to what felt like every doctor and nurse in the hospital. I couldn't imagine what else they could want to know that I hadn't already told them.

"I told you, Rogers," Dr. Patterson said, "we need permission from her parents to speak with her about this." He sounded upset.

Dr. Rogers ignored him. "We need to know the exact point that you first started noticing Mark's aberrant behavior."

I frowned "You mean, when he first acting crazy?"

"Well, when did you first notice him behaving— abnormally?"

"I thought we already covered this." I didn't mean to sound rude, but I was tired, and I *had* already answered this question about a million times.

Dr. Rogers consulted his clipboard. "So according to what you said earlier...the patient first started behaving abnormally when he saw the limo?" He glanced up at me. "Is that correct?"

I nodded, forcing myself to be calm. These doctors were here to help Mark. I should do whatever I could to make their job easier.

Dr. Patterson spoke up again. "Think carefully Susan. Before the incident with the limo, did you notice Mark acting— obsessively about anything else, or preoccupied with—uh success or winning, like winning a game or anything similar?"

I laughed. It took a few seconds for me to stop. I was so tired. "Mark always acts obsessive about some things, always a little O.C.D., if you know what I mean. But that's just who he is."

Dr. Patterson nodded, as if this was what he expected. He turned to the other two doctors. "Are you satisfied now?"

The doctor who hadn't spoken yet shook his head. "This doesn't prove anything Patterson. The symptoms can lay dormant for years. We're just now starting to understand how this thing manifests."

He stepped around Dr. Patterson, "Miss...Mitchell, is it? I'm Dr. Williams from the C.D.C., the Center for Disease Control."

I felt a knot form in my stomach. "Mark has some kind of— of disease?"

He cleared his throat and fidgeted with the papers on his clipboard. "Not exactly. We believe Mark may have been infected with a virus."

"A virus? You mean like the flu?"

"How long have you known Mark?"

I thought back to the first time we'd met, it seemed so long ago. "About four years—ever since he and his mother moved to our street."

"In that time, did you ever meet Mark's father?"

I shook my head, wondering where this was going.

"You and Mark spent quite a bit of time together though?"

I felt myself blushing. I was starting to not like Dr. Williams. "We were just friends."

"Though you did spend a lot of time together—as friends, I mean?"

I nodded.

"We'd like to take a small blood sample from you if that's OK, just to try to understand what's happening to Mark—and to make sure that you haven't been infected—if it is a virus, that is. It's merely a precaution."

Dr. Patterson stepped forward now, clearly agitated. "You cannot take a blood sample from a minor without her parent's consent!" he hissed.

Dr. Williams glared at him. "In the event of an epidemic, the C.D.C. has the authority to carry out testing of minors without parent—"

"There is no epidemic!" Dr. Patterson shouted over him. "All you have shown me is unrelated data about some moths and a few circumstantial cases from patients who—"

"Who have the exact same set of symptoms as this patient." Williams said, gesturing angrily towards Mark.

"What is going on?" a voice asked from the doorway.

We all turned. Mrs. Callahan stood in the doorway. My mother was behind her, looking at me anxiously.

"Are you this girl's mother?" Dr. Williams asked.

"I'm her mother," my mom said, stepping past Mrs. Callahan. "Why? She has a right to be here, we cleared it with the duty nurse."

"We'd like your permission," he said, shooting a quelling look at Dr. Patterson who had started to protest, "to take a blood sample from your daughter."

Mom looked from me to Dr. Williams, and I could see the concern in her eyes. "Why? What's wrong with Susan? Has she got—" she glanced at Mark, then at his mother, "Why do you need a blood sample?"

"There's a strong possibility that your daughter may be infected with the same virus that's responsible for Mark's condition."

~

Snow was falling outside my hospital window. I wondered if school had been canceled today. The snow didn't look bad yet, though it was hard to tell when you were lying in bed, strapped to a bunch of hospital monitors.

I'd been in the hospital for two days now. Well, two if you counted the time I'd spent as a visitor. That meant today was Wednesday. I was missing the Algebra review. It seemed odd to be worried about Algebra when I might be dying from some weird virus, but since I didn't feel sick, my concerns over Algebra took precedence.

I turned my head, careful to avoid disconnecting any wires. My mom was sitting in the only padded chair in our impromptu quarantine room. It was really just a regular hospital room, isolated from the rest of the building by a long hallway and several locked doors. Mrs. Callahan was still asleep. Mom had insisted that Mrs. Callahan take the room's only real bed, which meant that she and I got the flimsy cots that had been brought in. Still, I didn't complain. Mark's mom needed the bed a lot more than I did.

After the C.D.C. guy had gotten permission from my mom

to take my blood, he had also collected samples from my mother and Mrs. Callan. After that, he'd started grilling Mrs. Callahan about her husband.

Listening to her answers, I remembered that Mark had told me that he had very few memories of his father. He'd left when Mark was just a few years old, and his mother rarely talked about him.

By the time the doctors were finished with their questions, Mrs. Callahan was sobbing so hard she was almost incoherent. Between hearing that her son might be dying from some virus that made you crazy and having to relive the details of her husband running out on them, she was in pretty bad shape.

Then the doctors from the C.D.C told us we'd all have to stay in quarantine for the next few days until our test results came back, and she'd fainted.

The phone rang and mom picked it up. I hoped it was dad. Mom had called him right after we had been put in here. She said that he had been furious and was talking to a lawyer. I hoped that meant we'd be out of here soon.

"Yes, thank you for the news," Mom said. "We appreciate it." She hung up the phone and glanced over at Mrs. Callahan.

I was startled to see that her eyes were open. I guess she was a light sleeper, or maybe having your son in the hospital hard-wires your brain to wake you up when the phone rings.

"He's awake," mom told her with a tight smile. She looked as tired as I felt.

Mrs. Callahan sat up and squeezed her eyes shut. "Is he okay? Can I see him?"

"The nurse said until the doctor gives the order, we can't leave the room, but one of the doctors from the C.D.C. should be here soon."

Mrs. Callahan nodded, as tears rolled down her cheeks.

We all jumped as a sharp knock came at the door. The door

opened and Dr. Williams came in, followed by Dr. Patterson. Both of them were still wearing masks.

"How's Mark? Can I see him?" Mrs. Callahan started to get off the bed, but Dr. Williams held up his hand.

"Mark is stable. But until we're positive that we have a handle on the infection, we have to keep him in isolation."

"Speaking of infection," Dr. Patterson said, taking off his mask, "We have the results of your latest blood tests. You're all virus-free."

Dr. Williams scowled at him, but Dr. Patterson went on "You are all free to go. He smiled at Mrs. Callahan. "You may even go and see Mark in a few hours...after we're done with a few more tests."

Mrs. Callahan put her hand to her mouth, trying to choke back a sob.

"But is Mark—" I hesitated, glancing at Mrs. Callahan. "I mean—will he be okay now too?"

Dr. Patterson sighed. "We're just beginning to understand this virus and what its effects are. Now that we've started the antiviral treatments, Mark's immune system seems to have dealt with the virus quickly, which is fortunate given his—um, history."

"What history?" Mrs. Callahan asked. "Mark's never been sick like this before."

"This particular virus," Dr. Patterson said with a frown, "seems to only infect individuals with a specific genetic sequence. A certain pattern of DNA that makes them susceptible to its effects."

Mrs. Callahan looked pale. "Does this have something to do with Mark's father?" she asked, her voice barely a whisper.

"While we can't be sure, since we can't—um, find him," Dr. Patterson continued, looking uncomfortable, "we're fairly confident that Mark's father was infected with the same virus. Since half of Mark's genetic code comes from his father, that would make Mark a likely candidate for infection as well."

"But if you couldn't find him to do the test," Mrs. Callahan asked, "what makes you think he was infected?"

Dr. Patterson hesitated and Dr. Williams took the opportunity to jump in. "Because of the circumstances under which he left you and your son, Mrs. Callahan."

I scowled at him. I *really* didn't like this guy. Hadn't Mrs. Callahan been through enough already? Apparently, my mom felt the same way.

"I don't see what that has to do with anything," she said, her voice scornful.

Go mom.

"This particular virus," he went on, as if my mom wasn't giving him the stink eye, "is a new mutation of the baculovirus. The original strain only affected Gypsy Moth larvae. It caused them to climb to the tops of trees where they would remain until they died. Researchers speculate that this suicidal behavior was beneficial to the virus, because when the virus particles leave the body of the host, the high altitude allows them to spread for great distances, much farther than if the host had remained on the lower parts of the tree. This gives the virus a much better chance to infect new hosts."

"What has that got to do with Mark?" I asked, feeling annoyed. "He was trying to climb into a limo, not up some tree."

Thankfully, Dr. Patterson took charge again, so we were spared another lecture on moth larvae.

"As strange as it sounds," Dr. Patterson said, "the current theory is that like its counterpart in the moth, the virus that Mark contracted also wants to spread as widely as possible. Obviously, it benefits the virus if the host strives to socialize with as many people as possible."

"You mean the virus makes you want to be—popular?" I asked, bewildered.

"We aren't sure of its exact pathology," Dr. Patterson said, "but the strain that infects humans seems to compel its host to try to become increasingly more influential. The cases we're aware of so far indicate that the virus can cause extreme and irrational behavior, as long as that behavior would put the host into a better social position."

"Like—going to meet the governor in a limo," I said slowly.

"Or," Dr. Patterson said softly, looking at Mrs. Callahan tearstained face, "perhaps compelling a man to leave his wife and child so that he would have more time to excel at work."

Mrs. Callahan started sobbing again.

"It appears that just as the gypsy moth is compelled to become a tree climber," Dr. Patterson continued, "this virus compels men to become social climbers."

~

I looked out my kitchen window. The snow was falling harder than it had been earlier that morning. School had been canceled today, and I was glad. I wasn't ready to face the questions yet—or the homework I'd missed.

Mark was sitting across from me at the kitchen table, fidgeting with a button someone had left there. Same old Mark, always fidgeting.

"I can't stay long," he said, "My Mom has gone into super-over-protective mode since I've been home."

I smiled. "Can you blame her?"

"I guess not," he sighed. "It can be a bit annoying though. The doctors said I was fine. Officially virus free," he added, thumping his chest.

I made a face. "It's still so weird to think about. It almost seems like it was a dream."

"More like a nightmare," he muttered. He looked at me. "Look—I wanted to say I'm sorry for what I did. I—"

"Mark," I said, laying a hand on his arm, "it wasn't your fault, it was that—virus, not you."

He shook his head. "Part of me knows that's true, but the memories I have...it *felt* like me. Like the way I was acting was the way I *wanted* to act." He looked away from me, towards the window. "I remember wanting to get into that limo. I felt like—I knew—that I deserved to be there. That I needed to be more popular."

"Mark I—"

"Do you think that's really why he left?" He was on his feet now, staring out at the falling snow.

I knew he was talking about his dad. I wasn't sure what to say. Would my father ever leave us because of some virus? I shuddered.

"I'm going to find him," he said, his voice hard. "I'm going to find out what happened to him. I—I need to know the truth about why he left."

I watched Mark, thinking of his lists and little obsessions, and remembering his drive to succeed that helped us to win the school election. I couldn't help but wonder how much of his personality was really him, and how much was a result of the virus he'd been carrying all those years.

As he stood there, with that single-minded look he always got when he was determined to accomplish something, part of me wondered if he really was free of the virus.

∾

Principal Whittaker smiled as he walked out of his old office at Carter County Middle School for the last time. Earlier that day he had received word that after four long years of hard work and planning, he'd finally been appointed as the new Superintendent of Schools for Carter County. It was about time too. He knew that he deserved it. He'd been dreaming about this day for years. It was more than ambition—he needed this.

He needed to be popular.

Part Three
Frogs, Bats, and Wizards

The Science

———————— ⊗ ————————

I n the movie the Lorax, one of the characters proclaims, "When a guy does something stupid once, well, that's because he's guy. But if he does the same stupid thing twice, that's usually to impress some girl." If you've ever found this to be true in your life, you're not alone—it happens everywhere in nature.

Imagine that you're a frog. Not just any frog, but a mighty Túngara Frog from South America. Your species has enjoyed quite a bit of attention in the scientific press over the years, thanks to your lovely mating call. The researchers maintain that they are interested in you for purely scientific reasons, but you know the truth. Your mating call is simply irresistible to any species.

One evening, you're sitting in the forest, enjoying the warm evening breeze. Somewhere in the distance, a few crickets chirp feebly. Amateurs really, but one has to admire their willingness to try. As much as you enjoy sitting alone and basking in your own splendor, you decide that this evening you'd like some company, a special someone to share a moonlight stroll around the forest. So, you take a deep breath, open your mouth wide and—you're immediately interrupted by another frog's mating call from a few yards away.

You sputter and cough in indignation. Hadn't you clearly marked out this territory earlier today? Who is that ridiculous excuse for a frog that's poaching in your forest? Plus, his mating call sounded more like a strangled chicken instead of a mighty Túngara. You decide to show this interloper what a real mating call sounds like, you take a deep breath and—he interrupts you yet again!

Out of the corner of your eye, you catch a glimpse of movement at the edge of the clearing. Another frog, this one clearly female, has hopped into view. It's hard to tell from this distance, but she seems to be eyeing your competitor with interest and possibly—no, that couldn't be *admiration* in her eyes, could it?

This is outrageous. You decide to pull out all the stops, give that girl and her wimpy friend a demonstration of what a true Túngara mating call sounds like. Before you can be interrupted again, you take a deep breath and let loose the most amazing, elaborate, and beautiful mating call the forest has ever seen.

After a few seconds you stop, breathless. You glance up with a knowing smile, expecting, at the very least, that the female visitor will now have eyes only for you. Then you see something unbelievable. So unbelievable, you think at first, you're hallucinating due to a lack of oxygen. That girl is hopping away with the other guy.

Somehow, that pitiful excuse for a frog has wooed her with his wimpy mating call. If you had known anything about human music, you might have said it was like meeting a girl who preferred a man who could sing a barely passable Row, Row, Row Your Boat; instead of one who could sing the full chorus of Handel's Messiah. But of course, you're a frog and don't know anything about human music.

Dejected, embarrassed, and alone, you consider what to do

next. You can either admit defeat or, as your old father used to say when you were only a polliwog, "Son, the best lessons come from our defeats."

Of course, he was talking about the time he'd broken both his legs after discovering that frogs couldn't fly, but perhaps the same principle applied to mating calls.

Looking around timidly, you take a deep breath, swallow your pride, and give a much simpler mating call, one that sounds not unlike the call of the frog who'd stolen your potential girlfriend a few minutes ago.

Suddenly, you hear a rustling behind you. You can't believe it worked. Girls really did prefer a simpler mating call. But why was that rustling sound coming from above you—in the tree branches?

You're just about to hop away and call the night a loss, when a shadow swoops down from the treetops. With horror you realize it's not a frog at all, it's a frog-eating bat. You turn to flee, but you're too late. What started as a night filled with promise has ended in tragedy.

The astute reader will have realized from this account that the female Túngara frog preferred the frog with the less elaborate mating call. As with all the deep mysteries in life, the question of how female frogs choose their mates demanded an answer. So, a group of scientists from the university of Texas set out to understand this behavior a little better.

A scientist from the University of Texas named Mike Ryan gathered up some of the top frog-behavior experts in the field: a neuroscientist named Hamilton Ferris, and an ecologist named Amanda Lea.

But as Mike sat down to fill out the grant application necessary to fund this research, he realized that their team was still missing something. They needed someone who could really get inside the brain of frogs and the bats who ate them.

Enter Karin Akre, a scientist who describes herself as a biopsychologist, one of a select few who have training in both biology and psychology. Initially, the university system designated people with such training as psycho-biologists, but some worried that the title might call their sanity into question. So, in an effort to prove that psycho-biologists were just as sane as the rest of us, Karin agreed to join up with Mike and his team, and help unravel the mysteries of frog mating calls.

In their work, they discovered that female frogs did prefer complex mating calls to simple ones, but only to a point. Just like with humans, every girl frog has her limits. There's a threshold of quirky behavior she's willing to put up with in a potential mate before she just starts to ignore him completely.

The scientists found that female frogs preferred mating calls that were just slightly complex compared to really elaborate ones. At first, one might suspect that only frogs and psycho-biologists could tell the difference between a slightly complex frog mating call and a really elaborate one. But it turns out that bats can as well.

To show this, the scientists got a bunch of frog-eating bats together and tested to see which type of mating call was more likely to cause the bats to attack. Fortunately for the frogs, the scientists used recordings of mating calls rather than actual frogs in this experiment, but the results were as they expected. Just like the female frogs, the frog-eating bats preferred to hunt frogs with slightly complex, but not *too* complex, mating calls.

Rumor has it that these same scientists are planning a follow-up study to see if the reasoning for this is simply that frogs with simpler mating calls taste better. Readers interested in volunteering for this study should contact the scientists directly.

The takeaway message of this research is that frogs with a slightly complex mating call are more likely to find a mate than those with a highly elaborate one. This means that over the generations, the number of frogs willing to invest the time and money in learning how to create complicated mating calls has dropped off significantly.

The fact that female preference can drive the evolution of male behavior over time shouldn't really surprise anyone. In fact, this same principle is the reason why there are no more magicians in the world.

The Fiction

argret sat on one of the stone benches lining the spacious palace gardens and wondered why Gawain had to make everything so complicated. Their evening together had started out simple enough. Her father had agreed—albeit reluctantly, to let her take a moonlight stroll with Gawain. This close to summer, the air was still warm and the numerous palace guards provided ample supervision.

She tried to look interested in Gawain's display, but her eyes kept darting to the moon. They'd been gone for over an hour now. Her father was a patient man—but not *that* patient.

"Watch—this—one," Gawain panted, spinning his arms wildly.

She tried to look interested. But it was hard to focus on the colored lights swirling through the air while Gawain was flailing around below them like a chicken with its head cut off.

Magic, and by extension—magicians, had always fascinated Margaret. She'd been the envy of her sisters when Gawain had started courting her. But she'd been having second thoughts about their relationship for a while now—especially since she'd met William.

Margaret jumped as Gawain summersaulted into the air, squealing like a cat whose tail had just been stepped on. As he fell, a bright light streamed from his fingertips, coalescing into some kind of flying creature—a bird maybe? It swooped down and hovered just in front of her. She cringed as she realized it wasn't a bird—it was a bat. Why was magic always so creepy?

Gawain collapsed to the ground beside her, breathing heavily. As he lay on the ground, gasping for air, the magical light faded, leaving the two of them alone in the moonlight.

"Gawain," she said gently, "I really appreciate these—um, shows, but you needn't kill yourself just to impress me."

"No—trouble—my flower," he said between gasps. "You—deserve—the best."

Margaret felt herself blush and looked away uncomfortably. Should she tell him now? She'd better get it over with. Waiting longer would just make it worse. William had been patient enough with her continued friendship with Gawain, but she knew it bothered him that the two of them spent so much time together.

"Gawain...I wanted to talk to you about your um—proposal." She hoped the poor lighting would hide her embarrassment. Not that he would have noticed, she thought ruefully. Gawain never noticed anything when he was immersed in his magic. That was part of the problem.

Gawain sat up and flashed her a smug grin. "Yes...I'm listening."

Eyeing his confident expression, she wondered if he had ever tried to use magic to influence her feelings for him. No, she would have noticed if he had suddenly gone into convulsions while they were out together. She took a deep breath.

"Gawain I am flattered that you think so well of me, and I appreciate the gifts and magical displays, but I—"

"Which one was it?" he interrupted.

"I—what?"

"Which display was it that most impressed you?" he asked, his eyes intense. "I bet it was the time I turned the castle moat into ice, or maybe when I transmuted your horse into a frog."

Margaret winced. Starlight had never been quite the same after that. The next time the groomsman had tried to give the horse a bath, Starlight had become so spooked, she knocked the poor man into the water barrel. When she tried to explain to him why the horse had suddenly become terrified of water, the groomsman had muttered something about "worthless magicians" and stalked off to find a towel.

"You're right," Gawain said, misinterpreting her expression like he always did. "I suppose the horse transmutation wasn't as impressive as some of the others...like the time I made it snow—in the middle of summer!" He roared with laughter at his own cleverness.

"Actually," she said, remembering the protests of the farmers who had panicked at what had appeared to be an early freeze, "my favorite was when you made that diary for me."

He waved a hand dismissively. "Any junior apprentice can cast a bookbinding spell, Margaret. *I* can do much more complex magic than that. This one for instance—"

He started to rise, but Margaret grabbed his arm and pulled him back down. "Gawain," she said pointedly, "we're talking now, remember?"

"Sorry," he muttered. "I just thought you'd want to see this new spell I've been working on."

"Maybe later," she sighed, wondering again why he always had to make things so complicated.

"Gawain, listen. I like you a lot—but as a friend."

At those last words, his smile slipped a little. Then he seemed to regain his confidence.

"You know what they say, best friends make the best lovers."

"That's just it," she hesitated, bracing herself, "you're *not* my best friend Gawain, not anymore. And I'm—not in love with you." The last part came out in a rush.

He stared at her as if she'd suddenly started speaking another language.

"What are you saying? Do you mean you—" he stammered. "You mean to say that—are you turning me down?"

"Yes," she said, cringing slightly at the hurt in his face. "It's not you," she added hurriedly. "It's me. I'm just not right for you."

"You're—not right...for me?" He asked incredulously.

"I think magic is—um, interesting and everything," she said quickly, "But I just can't picture myself as the wife of a magician." She shuddered, remembering all the creepy magical stuff she's seen the last time they'd visited his workshop together.

"Is that all?" he laughed, sounding relieved. "I can fix that. What is it exactly? Does magic seem too difficult or taxing to you? I have a spell that can give you more energy—just when you're starting out—"

"Gawain, no. I—"

"One time," he said, cutting her off, "I used a spell to keep myself awake for three days straight so I could study for my master magician exams." He eyed her critically. "I think Antioch's Variation would be the best one for someone your size..." He started to stand again.

"No!" She pulled him down more forcefully this time.

He sat down with a thump and stared at her, blinking.

"That is exactly what I mean. You have to make everything so complicated. Everything for you is just one more problem to be solved with magic."

He looked at her like she was an idiot. It was his *I can't believe you don't know something this obvious* look. She hated that look.

"Margaret," he laughed. "Every problem can be solved with mag—"

"No! I don't *want* my problems solved with magic."

He stared at her, nonplussed.

"I just want...," she trailed off, her hand reaching subconsciously for the small iron band that she wore on her necklace—the band that William had given her after their first date. She fought back a smile, not wanting to offend Gawain more than necessary. "I just want a simple life Gawain."

Gawain looked at her, his eyes flicked to the necklace then back to her face. A look of cold understanding dawned in his eyes. "It's him, isn't it?" he asked in disbelief. "You're choosing some troglodyte blacksmith's apprentice over me."

Margaret turned red. "Don't you call him—"

"Me!" He was on his feet now, pacing. "One of the greatest magicians in the kingdom! Any other girl would be honored...to...to..." he spluttered.

"To what?" she countered, leaping to her feet. She hated it when he insulted William. "To watch you prance around like some demented rooster, just so you can make colored lights appear in the sky? In case you haven't noticed," she added, her voice full of contempt, "we have plenty of lights in the sky already."

Even in the dim light she could see his face turn red. "Is that what I am to you then? Just some...some...clown? Just some joke to be laughed at and then cast aside when you're done with me?"

Margaret took a deep breath, annoyed that she'd let herself get so angry with him. She knew that this was as much her fault as his. If she had ended it sooner, like her mother had suggested...but it was so hard to get him to listen to her. He always made it so complicated. He—

"And what about your parents? Are they going to be content

when they find out their daughter—the princess—is going to marry some soot-covered craftsman?"

"My parents don't care whom I marry Gawain," she said coolly, fighting to keep her temper in check. "My father has always made it clear that he had no plans to force any of his daughters into political marriages. He always promised we'd marry for love or not at all."

"*Love*," he sneered. "So, you're in love with this—this—"

"William. His name is William," she said stepping forward, her eyes hard. "Yes, Gawain. I'm in love with William. He's a good man, and unlike you, he actually cares about other people."

Gawain took a step backwards at the ferocity in her voice, then narrowed his eyes. "So...the princess and her blacksmith are too good for me then. I'll show you what you're losing out on."

"What going on?" a voice growled from the other end of the gardens.

Margaret turned and felt relief wash over her as William strode quickly towards them. He placed himself firmly between Margret and Gawain. He was still wearing his leather apron, and he was breathing hard. He must have come straight from the forges. Had her father sent for him?

"Oh goodie," Gawain muttered. "Look who crawled out of the soot pile...Prince Charming himself."

"William, I—" Margaret began, flustered.

William scowled at the Gawain then turned towards Margaret, his eyes softening. "Are you okay?"

"Of course she's fine, you simpleton," Gawain said. "She's with me, isn't she? We were both just fine until you showed up."

William narrowed his eyes and took a step towards Gawain. The massive blacksmith towered over him, and Gawain had to take a quick step backwards to keep looking him in the eye. Margaret tried not to smile.

"I think it's time for you to go back to your tower, magician," William said in a low voice.

"You think you can order me around, soot boy?" Gawain's eyes flashed, giving his expression an other-worldly glow. "Margaret is mine."

William's hands balled into fists. "Margaret," he said through clenched teeth, "is her own person. She doesn't belong to you or anyone else."

"Margaret would be lost without me," Gawain spat.

"What!" Margaret shrieked. "You conceited little—" She tried to step around William to slap the smug smile off Gawain's face, but William held her back easily with one powerful arm.

"You should go. Now," William said to Gawain. "Before you get hurt."

Gawain laughed nervously and took another step backwards. "You think some lowly blacksmith's apprentice is any match for me, the greatest magician in the kingdom?" He began twirling his hands above his head, his face set in a mix of concentration and anger.

Margaret tensed and latched onto William; her eyes closed tight against whatever horror Gawain was conjuring.

Crack!

Margaret opened her eyes hesitantly, expecting the worst. Instead, she saw Gawain sitting on the ground rubbing his jaw. He was staring at William's fist with wide eyes.

"Don't you ever," William snarled through clenched teeth, "try to use one of your filthy spells on Margaret or me again. Now leave—before I lose my temper."

Gawain got slowly to his feet, his eyes burning with hate. Without another word, he turned his back on them and stalked away.

Margaret felt her eyes sting as she watched him leave the palace gardens. They could have all been friends. She'd ruined everything. If only she'd said something sooner. Why did Gawain have to make everything so complicated?

William blew out a breath and shook his head. "I'm sorry," he muttered, turning to Margaret. "Are you okay? I shouldn't have done that. It's just—well when he started talking about you like you were some kind of—like he owned you, I—"

Margaret put a finger to his lips, silencing him. She looked up into those tender, sea-green eyes she loved so much. "It's my fault William. I should have handled that better. I should have spoken to him about our relationship sooner. I should have told him that you're the one that I—" she broke off, embarrassed.

William smiled and kissed her.

❧

The tradesman tapped his fingers impatiently on the wooden counter of Marcus' shop. Marcus noted with disapproval how dusty that counter had become lately. It had been a difficult few months for the bookseller and his family. But now that he'd made the changes his wife had suggested, things were starting to look up again. Still, he wanted desperately to sell off his old inventory, maybe make a little profit on it—or at least break even.

He reached behind his counter and grabbed another stack of books to show the tradesman. Marcus made a show of stacking the three volumes on the counter, arranging them so the man could see that the set was complete.

Marcus scratched his chin, pretending to deliberate. He didn't really need to deliberate—he'd tried unsuccessfully to sell these books five times this month. "A king's ransom is what I paid for 'em," he said thoughtfully. "But since you and I have been doing business for so long, I'll let ya' have the set for fifteen crowns."

"Sorry friend," the tradesman said, shaking his head. "I have too hard a time selling stuff like this anywhere. Every village between here to the eastern shore is still in an upheaval after last summer."

"Don't see where that changes nothin'," Marcus grumbled. It was a story he'd heard too many times. "Them's still good spell books. The whole series of Northrop's Arcane Alchemies and all." He eyed the man for a moment. "Twelve crowns then."

"I'm sorry Marcus," the tradesman smiled sympathetically. "I just can't sell them. Nobody's buying magic books anymore. Do you think it's just around the castle that all the young men have suddenly started trying to apprentice themselves off to craftsmen? Every girl in the kingdom has heard about Princess Margaret's wedding."

"S'just a fluke," Marcus grumbled, eyeing the books with dismay. Nobody had bought a spell book from his shop all summer. From what he'd heard from the other traders, it was much the same story all over the kingdom. People's interest in magic seemed to have suddenly dried up.

Fortunately, his wife had somehow understood that this drop in sales was more than a passing streak of bad luck. She'd encouraged him to start selling more mundane offerings— cookbooks, gardening tomes, and the like. It wasn't as lucrative as the spell book trade, but it kept food on the table.

"Ye just wait an' see," Marcus said. "This'll pass like things always do. Yer missin' out on a good bargain. It's still every boy's dream ya' know—t'go and become a magician."

The tradesman shook his head as he fastened his traveling cloak around his shoulders. "Not anymore my friend," he said, smiling ruefully. "Not when every girl's dream is to marry a blacksmith."

Part Four
Doors and Memories

The Science

Have you ever walked into a room to get something, only to immediately forget why you went in there? If this happens to you frequently, don't worry—you're not crazy. It turns out your brain is designed to do this. Scientists have studied this idea for some time now and have come up with some interesting theories to describe what's going on.

The human brain is a big spongy mass of gray goo all linked together with little electrical connections. Current brain research tells us that there are between fifteen and thirty billion neurons that make up a brain. Each of these neurons is connected to thousands of its neuron counterparts. When scientists have to deal with something that complicated, they don't bother trying to understand how each piece works. Instead, they develop a simplified model, which is often quite good enough to get on with. You use models all the time, whether you realize it or not.

For example, consider a light switch. When you turn on a light switch, a thin sliver of metal is moved around a pivot that connects to other pieces of metal. This connection allows a current of electricity to flow through the wires that run between the switch and the light fixture. That electrical current is generated at some power plant far away. This is most likely accomplished by burning a fossil fuel, which heats water enough to turn it into steam. The steam turns a turbine, which is connected to an electromagnetic generator. The generator produces an electrical current that passes through various transformers on its way to your house. As complicated as that sounds, even that isn't the whole story.

Chances are, unless you're a technician working on the power grid, you probably don't think about any of that when you stumble bleary-eyed into the kitchen first thing in the morning. The "model" most of us use for how the electrical system works is considerably simplified: turn on switch, light comes on.

The thing to remember about models is that they're always wrong—at least a little bit. Scientists know this, so they're always trying to come up with new experiments to refine their models and make them better.

Our simplified light switch model works perfectly most of the time. But eventually there comes a day when we discover our model is wrong. We turn on the switch, but the light doesn't come on. Now we have to refine our model to include the fact that there's a filament in a light bulb that might be burned out, and a circuit breaker in our basement that might be tripped, or a power line outside that might have been knocked out by a storm.

This is how science works—we build models to explain what we see happen in response to certain events, then when we find cases where our model doesn't explain what's going on, we come up with experiments to make the model better.

According to one model of how our brains work, our thoughts and memories get grouped together into events, and those events are then associated with contexts.

For example, when we're in the kitchen context, our thoughts might be engrossed by some recipe we're working on.

Then when we step into the living where a movie is playing, our brain undergoes a context shift. Those recipe thoughts get moved into the back of our minds to make room for the thoughts we need to understand what is happening in the movie.

Some scientists at the University of Notre Dame wanted to refine this model further. One of the things they discovered is that when we pass through a doorway, it triggers one of these shifts in context, causing our brain to file away memories from the previous room so it can make room for thoughts needed in the new room. They also found that just walking across a large room did not have the same memory effect that passing through a doorway between two smaller rooms did.

This idea of shifting memories made me wonder what would happen if someone could choose to shift a group of memories out of the way to make room for other memories. Even more interestingly, what would happen if you could use this technique to exchange a group of your memories with someone else's memories?

The Fiction

⟡

The cold, night air cut like icy daggers through Catherine's prisoner uniform as she crouched on the narrow window ledge above the deserted alleyway. There was a full moon tonight, but thankfully it was cloudy enough that the moonlight wouldn't betray her. Something cried out in the distance, though Catherine couldn't be sure if it was human.

She held her breath as two men that were no longer men passed directly below her. If they were following standard trapper procedure, they would continue along the street to the next intersection, then head left, completing the search pattern they'd been following all night. So far, only trappers had been sent after her. That was fortunate—she wouldn't have been able to avoid a coordinator by hiding on a window ledge.

The silent creatures shuffled down the narrow street below her, their jet-black clothing seeming to blend into the darkness completely. If it weren't for their chalky white skin, Catherine would have had a hard time seeing them—even from this distance.

She tensed as they paused at the end of the alleyway, then relaxed when they turned left and resumed their eerie shuffle down a side street. Catherine counted to ten before dropping down to the now deserted alleyway, then darted back the way the trappers had come—towards the facility.

Towards the facility. It sounded like madness. But if what she had learned was true—if Glen were still alive, then what choice did she have? To think she'd been *so* close—so close to freeing them both. No, she couldn't blame herself. There was no way she could have known that Glen was being held in the same facility she had escaped from. If she hadn't come across that guard...she repressed a shudder, trying not think how close she had come to losing Glen forever.

She crept further along the street, keeping to the inky shadows cast by the neglected buildings. Nobody lived in this part of the city anymore, not with the facility so close. In many ways, it was much harder to stay hidden in an abandoned city like this one. Her footsteps and breathing were the only sounds she could hear, unless you counted the beating of her heart, which was the only sound a coordinator would listen for.

Catherine knew from the information she'd shifted from the guard outside her cell that there should be one more pair of trappers coming exactly five minutes behind the first. She quickened her step, wanting to reach the drain before they caught up to her. She could probably evade the second pair just as easily as she had the first, but there was always a chance they would spot her, and then...she fought back the fear that threatened to overwhelm her. She couldn't think about what would happen if they caught her again. She had to get to Glen.

Catherine breathed a sigh of relief as she reached the pile of trash containers obscuring the opening to the drainpipe. The stench from the refuse made her want to retch, and she tried not to think about what might be in those containers, or how long they'd been sitting there. As she slid into the tight space between the grungy brick wall and the containers, she heard the faint sound of shuffling footsteps coming from the far end of the street.

"Right on time," she muttered, sliding the rest of the way into the slimy drainpipe. Her instincts told her to run, to get as far down that pipe as she could before the men that were no longer men passed too close. But she knew the pipe would amplify any noises she made. Trappers weren't very clever, but they had excellent senses.

She smiled wryly to herself as she remembered the stories her friends used to tell about the Syndicate trappers back in the camps. How they could track anyone, find anyone just by thinking of their name. The stories said that no one could hide from a trapper for long. It had been a long time since Catherine had been that naive. They *did* have a very keen sense of smell. Once they had your scent, it was almost impossible to shake them.

Catherine had shifted that little tidbit of knowledge from a Syndicate soldier a few years ago, just before the man had died. It had been close, shifting memories from the dying was risky, but this risk had been worth it. This information had saved her life.

The trappers passed her as silently as the first set had done. As repulsive as the pipe smelled, Catherine knew it would mask her scent well. She waited a few seconds as the trappers continued up the street, following the same circuit as the first group. Once they'd passed out of sight, she slowly turned and began inching her way along the dark tunnel.

～

Glen stared at the stark white walls of his cell and silently cursed the irony of the universe. Just three days ago he'd been trying to get Catherine to leave him. He'd tried, unsuccessfully, to persuade her that she would be safer with the northern resistance camp. It was closer to the fringes of Syndicate influence and saw considerably less fighting.

But of course, the stubborn girl had adamantly refused to leave. Finally, he had managed to arrange a ruse with their camp commander to make it look like they were both being transferred to the northern camp. His plan was to see her there safely, then he would receive a new set of orders requiring him to return to the southern camp alone. Catherine would be safe, and he'd keep fighting the Syndicate where the resistance needed him most.

They had arrived at the northern camp without incident, and he was just congratulating himself on his brilliant plan when a full squad of Syndicate collectors descended on the camp, plunging his world into chaos. He'd lost track of Catherine during the fighting, but hoped she'd had enough sense to escape. He shook his head angrily, knowing that she wouldn't have even tried to escape—she would have tried to find him.

Glen raised his head at the sound of some kind of commotion in the hallway beyond his cell door. He crossed the room, leaned close to the door, pressing his ear against it as he tried to make sense of the sounds. There were footsteps—booted footsteps—running. Someone was shouting, it sounded like hurried orders, though he couldn't make out the words. Was there some kind of prison break going on? He'd never heard of a Syndicate prison break before, at least not one where any of the prisoners had survived.

He jumped as he heard someone activating the security panel. He crouched next to the door, ready to pounce on whoever entered. If there was a prison break, he wanted in, even if it was suicide. Anything was better than sitting around waiting for whatever the Syndicate had planned for him.

The door slid open, and he sprang for the guard's knees. The guard sidestepped nimbly out of the way at the last minute, and Glen's momentum sent him crashing painfully into the wall.

"Glen, wait!"

He spun around and gasped. Catherine stood in front of him, wearing a slightly oversized Syndicate guard uniform.

~

"But how did you get the security code to the door?" Glen panted as he hurried to keep up with her. Catherine held up a hand as she peered around the corner, then motioned for him to follow.

"And how exactly do you know where we're going?" he hissed quietly as Catherine led him through the maze of facility corridors.

She had considered taking a different path back to the drain she'd used to sneak in, afraid that the Syndicate would have figured out by now how she had escaped the first time. But the memories she'd shifted from the duty officer told her they still hadn't discovered the broken storm grate that led to the escape tunnel.

She glanced back at Glen, saw the wary look in his eyes and sighed. Showing up in a Syndicate uniform had been a bad idea. Glen distrusted everything to do with the Syndicate, everyone in the resistance did. What would he say if he found out what she could do?

"Let's focus on getting out of here first," she said. "I promise I'll tell you everything when we're safe."

He held her gaze for a moment, his eyes narrowing. Finally, he gave a quick nod. "Let's go then."

❧

"Let me get this straight," Glen said. He took a deep breath and tried to keep his voice level. "You admit that you're a shifter, but I'm supposed to believe that you never worked for the Syndicate?"

Catherine nodded slowly, refusing to look away. She was seated on one of the two dusty benches in the small building they had decided to hide in. It was a good hiding place, close to the outskirts of the city, but not so isolated as to stand out.

She watched Glen pace back and forth, running his hands through his hair in agitation. Catherine was long past feeling ashamed of what she was, and she knew that showing weakness in front of Glen would only make him more suspicious. So, she had told him everything. At least, she'd told him everything she could still remember. She desperately wanted him to believe, to trust her again, but part of her knew that was impossible now.

Glen blew out a breath and sat down on the bench across from hers. It pained her to realize how much she'd been hoping he'd sit next to her. She had hoped for it, but didn't expect it. Still, Catherine knew the worst was over. He hadn't run. He hadn't abandoned her. He was still here—that was something.

Glenn squeezed his eyes shut and rubbed his temples. "Why didn't you tell me before?"

She snorted. "Glen, I just saved your life—after I was already safely away from the Syndicate. I risked recapture to go back for you. If after all that, you still don't trust me, what do you think your reaction would have been if I'd told you everything when we first met?"

He only grunted. His eyes were still shut tightly, but his fists had unclenched. He was close enough that if she extended her arm, she could touch his hand. She considered it but was afraid he'd pull away.

"Glen," she said softly. "I didn't ask for this. You can't imagine how much I hate it. How much I hate knowing that I was—bred," she shuddered, "to steal secrets and plant lies for the Syndicate. I—"

"How does it work?" he asked, his voice hard. "What happens to your—victims, when you're done with them."

Catherine cringed at his words, but tried not to let it bother her. What had been done to her wasn't her fault and she refused to feel guilty for it. "When I—when a shifter makes physical contact with someone, they have access to all their memories. But we can't just take a memory, we can only make an exchange."

He frowned. "What do you mean, *an exchange?*"

"Imagine a library of books, but the only way to pull one off the shelf is to replace it with a book you already own. That's how it works. Whenever I take a memory from someone, I have to replace it with one of mine. I—" her voice broke. "I lose a part of myself every time."

Glen opened his eyes. "Have you ever taken one of *my* memories Catherine?" he asked softly.

"No," she said firmly.

"Would I—I mean, would someone know when it happens? Can they tell that their memories have been stolen and replaced with one of yours?"

Catherine shook her head sadly. "They have no way to know. It's just like forgetting anything. You know that you knew it, but you just can't remember the details."

"And what about the memory that you exchange? Don't they realize that they have this odd, extra memory floating around in their head?"

"They usually don't notice it right away. The memory just sits dormant in their mind until something triggers it. The feeling they have is like déjà-vu, a memory of something you can't quite place."

Glen grunted again, then scowled down at the floor.

"The worst part," she said in a low voice, almost to herself, "is what we have to give up."

He looked up at her, confused. "What do you mean? I thought you said you choose to make the exchange."

"We choose to make the exchange, and we choose which memory to take, but we don't choose which of our memories we give up."

His eyes widened. "It's just—random?"

Catherine shook her head. "The ones from our early childhood are the first that we lose. I—" her voice broke again, and she took a deep breath to steady herself before continuing. "I know I had a family Glen—know that they loved me—but I can't remember their faces. I can't even remember their names."

References

Mice, Bacteria, and Goblins

Duan, F., & March, J. C. (2010). Engineered bacterial communication prevents Vibrio cholerae virulence in an infant mouse model. Proceedings of the National Academy of Sciences , 107 (25), 11260–11264. doi:10.1073/pnas.1001294107

Zombie Ants and Socialites

Hoover, K., Grove, M., Gardner, M., Hughes, D. P., McNeil, J., & Slavicek, J. (2011). A Gene for an Extended Phenotype. Science , 333 (6048), 1401. doi:10.1126/science.1209199

Frogs, Bats, and Wizards

Akre, K. L., Farris, H. E., Lea, A. M., Page, R. A., & Ryan, M. J. (2011). Signal perception in frogs and bats and the evolution of mating signals. Science (New York, N.Y.), 333(6043), 751–752. doi:10.1126/science.1205623

Doors and Memories

Radvansky, G. A., Krawietz, S. A., & Tamplin, A. K. (2011). Walking through doorways causes forgetting: Further explorations. The Quarterly Journal of Experimental Psychology, 64(8), 1632–1645. doi:10.1080/17470218.2011.571267

About the Author

LEE FALIN is a best-selling middle-grade and young adult author, and the host of the My Cousin Jane podcast. Lee has a PhD in Genetics, but these days he writes more fiction than research.

Lee is married to a wonderful woman he met while they were both serving as volunteer missionaries in Brazil. They have five amazing children, a shocking number of which are now adults.